INSIDE ANIMAL HOMES

Inside Anthills

Henry Abbott

PowerKiDS press.
New York

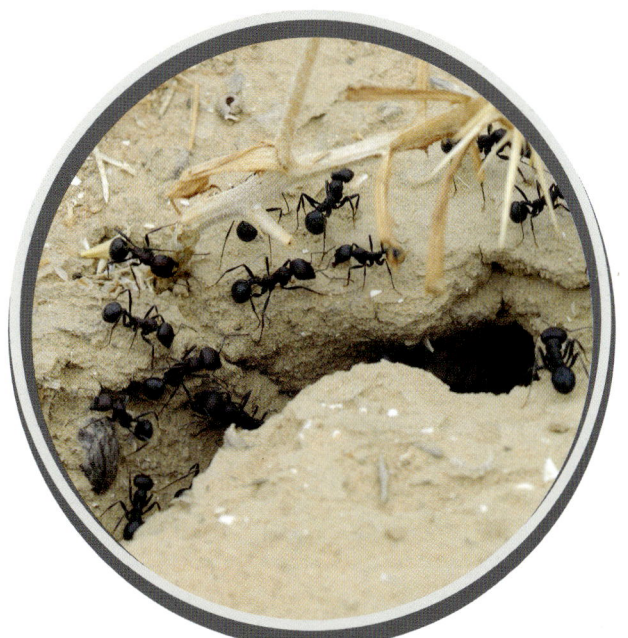

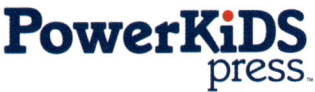

Published in 2016 by The Rosen Publishing Group, Inc.
29 East 21st Street, New York, NY 10010

Copyright © 2016 by The Rosen Publishing Group, Inc.

All rights reserved. No part of this book may be reproduced in any form without permission in writing from the publisher, except by a reviewer.

First Edition

Editor: Sarah Machajewski
Book Design: Mickey Harmon

Photo Credits: Cover, pp. 1, 5 (ants) AG-PHOTOS/Shutterstock.com; cover, pp. 3, 4, 6, 8, 10, 12, 14, 16, 18, 20, 22–24 (dirt) Raduga11/Shutterstock.com; cover, pp. 1, 3, 4, 6, 8, 10, 12, 14, 16, 18, 20, 22–24 (magnifying glass shape) musicman/Shutterstock.com; pp. 6, 8, 12, 16, 18 (magnifying glass) tuulijumala/Shutterstock.com; p. 7 (background) Marquisphoto/Shutterstock.com; p. 7 (ant) Henrik Larsson/Shutterstock.com; p. 9 successo images/Shutterstock.com; p. 11 S1001/Shutterstock.com; p. 15 Visuals Unlimited, Inc./Joe McDonald/Visuals Unlimited/Getty Images; p. 17 Mitsuaki Iwago/Minden Pictures/Getty Images; p. 19 SCOTT CAMAZINE/Science Source/Getty Images; p. 21 belizar/Shutterstock.com; p. 22 Hamik/Shutterstock.com.

Library of Congress Cataloging-in-Publication Data

Abbott, Henry, author.
 Inside anthills / Henry Abbott.
     pages cm. — (Inside animal homes)
 Includes bibliographical references and index.
 ISBN 978-1-4994-0860-7 (pbk.)
 ISBN 978-1-4994-0883-6 (6 pack)
 ISBN 978-1-4994-0909-3 (library binding)
 1. Ants—Juvenile literature. 2. Animals—Habitations—Juvenile literature. I. Title. II. Series: Inside animal homes.
 QL568.F7A36 2015
 595.79'61782—dc23
                                    2014049575

Manufactured in the United States of America

CPSIA Compliance Information: Batch #WS15PK: For Further Information contact Rosen Publishing, New York, New York at 1-800-237-9932

Contents

Home Sweet Home............................4
Buggy Body....................................6
More Than Meets the Eye..................8
Ant Nest......................................10
Tunnels and Chambers....................12
Ant Architects..............................14
Storing Food................................16
The Queen Ant.............................18
Staying Safe................................20
Ant Antics...................................22
Glossary.....................................23
Index...24
Websites....................................24

Home Sweet Home

What is your home like? How many people live there? Now, imagine building your home in the dirt or a tree and inviting more than a million people to live with you! This may seem strange for people, but it's not strange for ants. Ants live in **colonies** just like this.

Ants are one of the hardest-working bugs. They work to keep their colony healthy and strong. This all begins with having a place to live. Let's go inside anthills to see how they help this bug **survive**.

This hole is a doorway to an ant home.

Buggy Body

Ants are common bugs. There are more than 10,000 kinds of them. They live all over the world, except in very cold areas. In some places, almost half of all the bugs living there may be ants.

An ant's body has three parts. The first part is a large head. The middle part is the **thorax**. The third part is the **abdomen**. The thorax and abdomen are joined by a narrow part called a waist, which is something ants are known for.

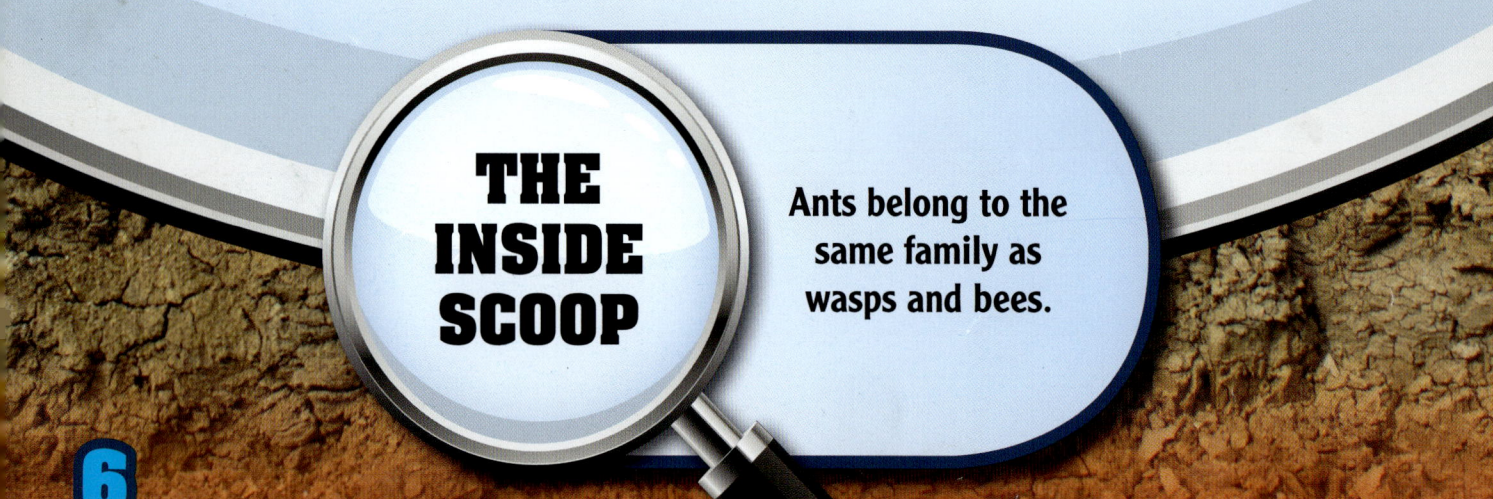

THE INSIDE SCOOP

Ants belong to the same family as wasps and bees.

Ants have six legs. They also have feelers on their head.

More Than Meets the Eye

Ants are very social bugs. They live and work together in large colonies. Ant colonies are highly **organized** communities. Each ant helps the community survive.

The biggest ant colonies have millions of members. Ants have to build a home that's big enough for everyone. Often, the place with the most space is under the ground. Ant homes look simple from the outside, but there's more to them than meets the eye.

THE INSIDE SCOOP

There are three kinds of ants in a colony—female worker ants, male ants, and a queen. Males and queens have wings. Female workers don't.

This may look like a mound of dirt, but it's the doorway to a home that houses millions of ants.

Ant Nest

Ant homes are called nests. You may hear people call them "anthills." That's because the outside of a nest can look like a hill of dirt. The dirt comes from all the digging ants must do to build their nest.

Many ant nests are under the ground. Some kinds of ants build nests in trees. Ant nests can also be found under rocks and piles of leaves. Tiny ants can make nests inside acorns! No matter where it is, an ant nest has one job: to house the colony.

The more the ants dig, the taller an anthill gets.

Tunnels and Chambers

Let's go inside an ant nest that's been built under the ground. The opening in the anthill is called the entrance. Ants use this as a doorway to get in and out of their home.

The inside of an ant nest is an organized **network** of tunnels and chambers. The tunnels are like hallways ants use to travel to different areas inside their nest. The tunnels empty into chambers. Chambers are much like the different rooms in a house.

THE INSIDE SCOOP

Some ant nests have one entrance. Very large ant nests may have several.

This image shows what an ant nest looks like inside.

Ant Architects

Ants are some of nature's greatest builders. It takes a lot of skill to plan such an organized home. They build their home by biting and digging.

Ants have two pairs of **jaws**. One pair helps them bite and carry. They bite off tiny bits of soil and carry them out of the nest. Over days and weeks, this creates a network of tunnels and chambers. Ants learn how to travel around their nest by following scent trails left by other ants.

Ants can carry 10 to 50 times their own body weight.

Storing Food

Each chamber in an ant nest is different. Some are used to store food. Worker ants travel outside the nest to find food. They carry it back to their colony and drop it off in the chambers. When it gets cold or hard to find food, ants eat what they've stored.

Some worker ants store food inside their body. These ants eat a lot, and their abdomen gets very large. These ants get too big to leave the nest, so they stay in. They live in a chamber deep inside the nest.

THE INSIDE SCOOP

Worker ants that store food in their body are called honeypot ants. They fatten up on water and sweet liquids, such as honey!

When ants want to eat a honeypot ant's food, they rub them with their feelers. The honeypot ant throws up a little bit of liquid, which becomes dinner for the colony.

The Queen Ant

The queen ant is the most important ant in the colony. Her job is to lay eggs. This keeps the colony alive. Because the queen is so important, she hardly ever leaves the nest. Staying inside keeps her safe.

The queen lives in her own chamber. Male ants come together with her to make babies. The queen lays eggs, which get carried to other chambers by worker ants. There, worker ants tend to the eggs and ant **larvae** until they grow up.

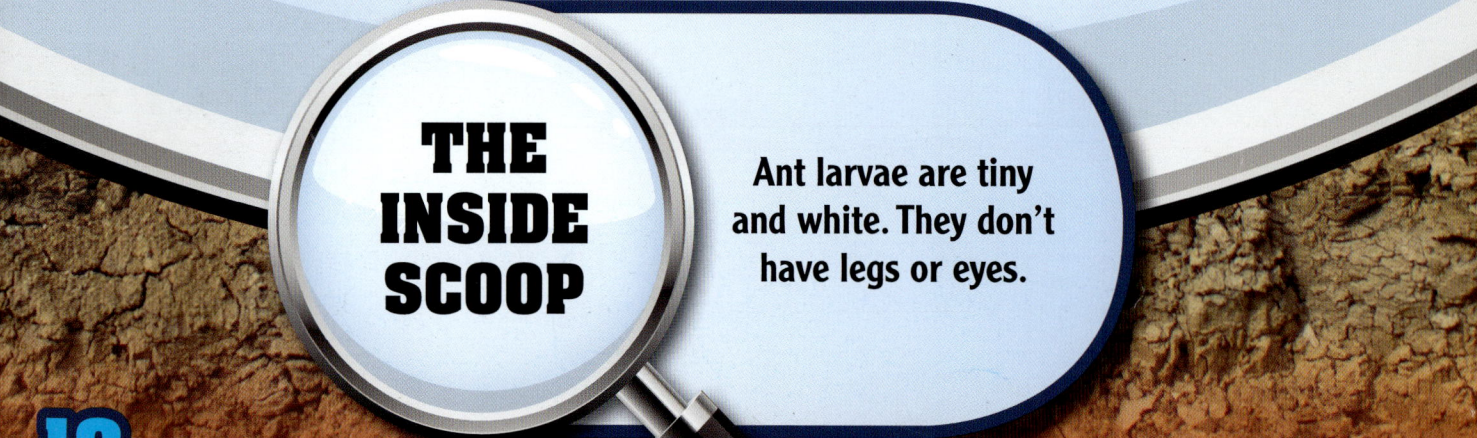

THE INSIDE SCOOP

Ant larvae are tiny and white. They don't have legs or eyes.

Queen ants are born with wings. They fly away from their nests to make new nests. Then they lose their wings.

Staying Safe

Ant nests give ants somewhere to live, store food, and grow their colony. They have one more important job—keeping ants safe.

Ants are food for many animals, such as anteaters. Anteaters use their sharp claws to tear apart nests. Then they use their long tongue to scoop up ants. An anteater's tongue may be long enough to reach 2 feet (0.6 m) inside the nest. Ants keep their queen, her eggs, and their food supplies safe by building those chambers deeper than predators can reach.

Ants **protect** their nest by attacking predators. All ants can bite, and some can sting. Some ants even spray **chemicals** to fight their enemies.

Ant Antics

With so many kinds of ants in the world, it's no surprise that they all live a little differently. Yellow crazy ants build **supercolonies** that have several queens. Army ants don't build nests at all. Instead, they move around and look for food. Carpenter ants dig tunnels in wood—including houses!

The next time you go outside, look around and see if you can find an anthill. It's just the beginning of something very cool below the ground.

Glossary

abdomen: The part of a bug's body that holds the stomach.

chemical: Matter an animal makes inside its body that can cause changes to its body or surroundings.

colony: A community of plants or animals.

jaw: One of the body parts that shapes an animal's mouth.

larvae: Bugs in an early life stage that have a wormlike form. The singular form is "larva."

network: A system in which all the parts are connected.

organized: Arranged in a highly advanced way.

protect: To keep safe.

supercolony: A large ant colony that houses millions of ants, including several queens.

survive: To live through something.

thorax: The part of a bug's body that holds the heart and lungs.

Index

A
abdomen, 6, 7, 16
anteaters, 20

C
chambers, 12, 13, 14, 16, 18, 20

E
eggs, 18, 20
entrance, 12, 13

F
feelers, 7, 17

H
head, 6, 7
honeypot ants, 16, 17

J
jaws, 14

L
larvae, 18
legs, 6, 7, 18

M
male ants, 8, 18

N
nests, 10, 12, 13, 14, 16, 18, 19, 20, 21, 22

P
predators, 20, 21

Q
queen, 8, 18, 19, 20

T
thorax, 6, 7

W
waist, 6, 7
wings, 8, 19
worker ants, 8, 16, 18

Websites

Due to the changing nature of Internet links, PowerKids Press has developed an online list of websites related to the subject of this book. This site is updated regularly. Please use this link to access the list: www.powerkidslinks.com/home/ants